AF521370

Raiganj Wildlife Sanctuary: A Safe Home for Migratory Birds

Prof. Chanchal Kumar Manna
Dr. Alok Kumar Pramanik
Dr. Kalyan Brata Santra

SCIENTIFIC® PUBLISHERS

Published by
SCIENTIFIC PUBLISHERS (INDIA)

Jodhpur –

5 A, New Pali Road
P.O. Box 91
Jodhpur - 342 001 INDIA

Delhi –

4806/24, Ansari Road
Daryaganj
New Delhi - 110 002 INDIA

Print: 2024

All rights reserved. No part of this publication or the information contained herein may be reproduced, adapted, abridged, translated, stored in a retrieval system, computer system, photographic or other systems or transmitted in any form or by any means, electronic, mechanical, by photocopying, recording or otherwise, without written prior permission from the publisher.

Disclaimer: Whereas every effort has been made to avoid errors and omissions, this publication is being sold on the understanding that neither the editors (or authors) nor the publishers nor the printers would be liable in any manner to any person either for an error or for an omission in this publication, or for any action to be taken on the basis of this work. Any inadvertent discrepancy noted may be brought to the attention of the publisher, for rectifying it in future editions, if published.

Trademark Notice: Product or corporate names may be trademarks or registered trademarks, and are used only for identification and explanation without intent to infringe.

© 2022, Publisher

ISBN : 978-93-92590-27-6
eISBN : 978-93-92590-29-0

Printed in India

Preface

Raiganj Wildlife Sanctuary is an artificial planted forest of the Forest Department, Govt. of West Bengal. The unique feature of this forest is that it provides the nesting place for many migratory and some resident birds species. Every year, a huge number of Asian open billed stork, Cormorant, Heron, Egret and other bird species used to come in this Sanctuary for their breeding purposes. The main attraction of various migratory birds' species is to find safe nesting zone, availability of nesting materials, suitable climatic conditions and plenty of food materials. No intraspecific and interspecific tussle for food materials are observed in this sanctuary.

In this book an attempt has been made basically on some ecological factors of the Sanctuary that influence the bird's breeding life and safety of their newborn individuals. The present observations have been made to study some aspects of breeding behavior, diversity in nest pattern and parental care of diverse migratory bird species in the Sanctuary. Many factors are associated for the purpose of breeding, nest site selection and day to day activities of various migratory species. The main objective of this study is to investigate the factors that affecting the bird's population in the forest area of the Raiganj Wild Life Sanctuary.

Raiganj Wildlife Sanctuary accommodates a good number of flora which attracts the various bird species (migratory and nonmigratory). The surrounding environment is very calm. A minimum types of disturbances were there which does not interfere with the normal life of the migratory bird species. In this book, special emphasis have been made to study the interaction between different flora and fauna with various migratory bird species. Considering the various environmental factors all the studies have been made within the same nesting zone of migratory bird species in the Raiganj Wildlife Sanctuary, Raiganj, Uttar Dinajpur, and West Bengal, India.

— Prof. Chanchal Kumar Manna

— Dr. Alok Kumar Pramanik

— Dr. Kalyan Brata Santra

ACKNOWLEDGEMENTS

The Authors of this book are grateful for the kind permission granted by the Department of Zoology, University of Kalyani, Nadia, West Bengal, India.

This type of work could not be done without the help and permission of the Authories of the concerned Departments, e.g., the Chief Conservator of Forests, Wildlife (Head quarters) and Chief Wildlife Warden, Government of West Bengal, India; Conservator of Forests, North-West circle, West Bengal; Divisional Forest Officer, Raiganj Division, West Bengal. We are really grateful to Mr. Anil Roy, Raiganj Division, West Bengal for issuing permission to carry out the present research work within the Raiganj Wildlife Sanctuary.

We are indebted to Dr. Premadhis Das, Reader in Statistics, University of Kalyani for statistical analysis and Scientists, Zoological Survey of India and Botanical Survey of India for helping in identification of various fauna and flora available in the Raigang Sanctuary . The Authors are grateful to Prof. G. G. Maity, Department of Botany, University of Kalyani for extending his expertise in identifying of various flora of Raigang bird Sanctuary.

Our sincere thanks are extended to Prof. B.Mondal and his colleagues , Soil Chemistry Laboratory, Bidhan Chandra Krishi Viswavidyalaya, Nadia, West Bengal for offering the soil testing facilites. Undoubtedly, there are many individuals and institutions not mentioned here but have been responsible for the successful completion of this work.

— Authors

About Author's

Prof. C.K. Manna

Prof. (Dr.) Chanchal Kumar Manna (Retd.) [*Professor in Zoology (Retd.), Department of Zoology*, University of Kalyani, Kalyani, Nadia, West Bengal, India] has been associated with teaching and research for more than 42 years in the college and University. He guided more than 23 Ph.D. students and published more than 100 scientific papers in the reputed journals. He also published books from India and abroad.

Dr. A. K. Pramanik

Dr. Alok Kumar Pramanik [*Department of Biological Sciences*, Nayabazar High School (H.S.), Nayabazar, Gangarampur, Dakshin Dinajpur, West Bengal, India] is currently engaged in teaching and research. At present he is conducting research work on the migratory birds in the Raiganj Wildlife Sanctuary. He has 10 research publications in various leading scientific journals and presented papers in various national and international seminars. He has published a book on migratory birds from India.

Dr. K. B. Santra

Dr. Kalyan Brata Santra [*Department of Biological Sciences, Jadupur Anchal High School (H.S.)*, Jadupur, Kamlabari, Malda, West Bengal, India] is carrying out research works on birds' ecology and Reproductive Biology. He has participated and presented papers in various national and international seminars. He has published more than 26 research articles in various scientific journals. At present he is engaged in teaching and research. He has also published a book from India.

CONTENTS

Preface *iii*

Acknowledgements *v*

About Author's *vii*

- **Raiganj Wildlife Sanctury** **1–4**
- **Various Ecological Factors of the Raiganj Wildlife Sanctuary** **5–6**
- **Major Flora and Faunal Diversity of the Raiganj Wildlife Sanctuary and their Interactions with the Migratory Birds Species** **7–45**
- **Factors affecting the Population of Asian Open-Billed Stork, The Major Migratory Bird in this Sanctuary** **46–57**
- **Conclusion** **58–63**
- **References** **64–71**

RAIGANJ WILDLIFE SANCTURY

India is one of the greatest and exclusive places on our planet that offers beautiful species of wild life especially when it comes to birds. India is home for approximately 1250 different species of birds living here in diverse climatic conditions, vegetations and habitats. From Himalayas to Kerala or Gujrat to North-East, every nook and corner of India offers scores of species of birds. And the best destinations to experience them are the nationwide spread bird sanctuaries. Indian bird sanctuaries are bestowed with flora and fauna. Raiganj Wildlife Sanctuary is one of the most notable bird sanctuary in India. This wildlife sanctuary popularly known as Kulik Bird Sanctuary, because the river Kulik flows beside the sanctuary and different migratory and residential bird species are the main fauna of the sanctuary.

The Raiganj wildlife Sanctuary is an artificial planted forest of West Bengal (Fig. 1), India. The area of the sanctuary is around 1.30 Km^2. The core area is about 0.14 Km^2 and the rest is the buffer area. The river Kulik flows around part of the sanctuary and acts as the boundary in its eastern and southern parts. The western part is marked by the farm of the State Sericulture Department. The shape of the sanctuary is that of the English alphabet "U". The Kulik river serves water to the sanctuary. Because the sanctuary has a network of artificial canals connected with the river Kulik. During monsoon the river water enters the sanctuary through the canals, which also supports a wide variety of food for the birds. The main and deepest canal runs through the middle of the sanctuary and forms a small island (Fig. 2) in the core area of the sanctuary. The water is retained througout the year and the depth of water is 2.5 m during monsoon. Apart from this canal, other canals have the water with 1.5 m depth and 0.O m depth water presents in low land area of the sanctuary during monsoon. In summer, there is no water except main canal (Fig. 3) and river Kulik. The authority of the sanctuary made more canals within the forest range for better management.

Raiganj Wildlife Sancuary belongs to the co-ordinates of 25° 37′ N to 25° O2′N and 88° 07′ E to 88° 12′E. It is situated near Raiganj in Uttar Dinajpur district (Fig. 1), West Bengal, India. It is located 4 km (2.5mi) north from the centre of Raiganj town, the district headquarter. National Highway 34 runs by its southern border and connects Raiganj with Kolkata and Siliguri. Raiganj is 425 km (2O4mi) form Kolkata and 181 km (112mi) from Siliguri by road.

Raiganj Wilidlife Sanctuary has an enriched wildlife that includes flora and fauna as well. The sanctuary is the home of many birds' species. The broad – leaved deciduous trees provide the nesting place of many migratory and resident birds' species. About 70,000 to 80,000 migratory birds visit the sanctuary every year for nesting purpose.

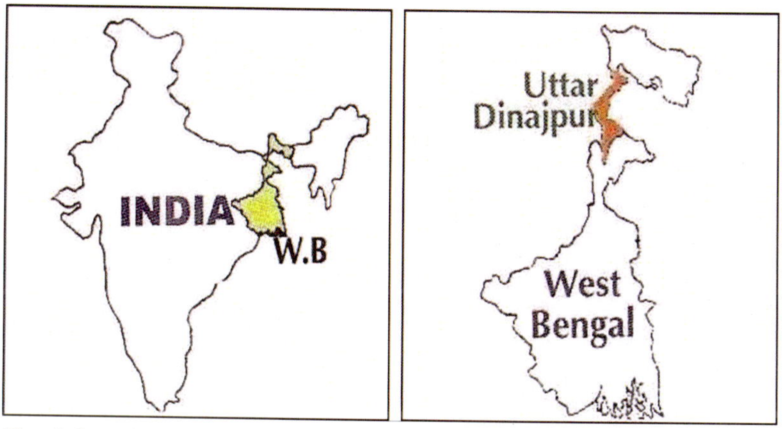

Fig. 1: Location of the Raiganj Wildlife Sanctuary.

Fig. 2: Horse-shoe shaped island of the Raiganj Wildlife Sanctuary surrounded by canal during winter.

Fig. 3: Horse-shoe shaped island of the Raiganj Wildlife Sanctuary surrounded by canal during summer.

VARIOUS ECOLOGICAL FACTORS OF THE RAIGANJ WILDLIFE SANCTUARY

In the Raiganj Wildlife Sanctuary, premonsoon (March-June) was characterized by high air temperature (average maximum temperature 2Q13 °C in 2007 and 2Q43 °C in 2008 in the month of June), low percentage of relative humidity (average 54.87 % in 2007 and 55.33 % in 2008 in the month of April) and occasional rain (average 0.800mm in 2007 and 0.774 mm in 2008 in the month of March). The monsoon season (July –September) had high percentage of relative humidity (average 82.87% in 2007 and 83.23% in 2008 in the month of September) and higher rainfall (average 14.333 mm in 2007 and 9.000 mm in 2008 in the month of September). On the other hand the postmonsoon season (October-February) was characterized by lower temperature (average minimum temperature 8.42 °C in 2007 and 9.74 °C in 2008 in the month of January) and lesser precipitation (average 0 mm in 2007 and 2008 in the month of December, January and February).

The pH of water of the sanctuary remained slightly acidic during monsoon season. Conductivity value was maximum during summer and minimum during winter months. BOD values have been recorded higher during premonsoon season. The minimum value of CO_2 was observed during winter months. The increase in total alkalinity during summer months appeared to be due to the concentration of nutrients in water. Total suspended solid and total dissolved solid values were maximum in summer and minimum in winter. Premonsoon showed high chloride values when water level was low followed by monsoon and postmonsoon with high water levels. Nitrate value was higher during monsoon season and decreased in postmonsoon and premonsoon season. High amount of phosphate remained in the canals during summer months.

The pH values of soil were recorded Q0 and below Q0. The nature of soil of the sanctuary seemed to be acidic. Percentage of organic carbon increased after premonsoon season in both the core and buffer region and this value became highest during postmonsoon season. High concentration of available nitrogen and available potash was observed in the core region as well as buffer region during postmonsoon than premonsoon season. Availlable phosphate content value was lower than the nitrogen and potash content of soil of the sanctuary. The variation of soil characteristics may be influenced by faeces of birds.

The meteorological data revealed that climate of the Raiganj Wildlife Sanctuary is characterized by a distinct seasonality of rainfall, dry summer period and comparatively cooler winter season. The physico-chemical characteristics of water expressed the poor water quality in the canals of the sanctuary. This characteristic showed maximum contamination of water due to faeces of birds. The soil of Raiganj Wildlife Sanctuary is mostly acidic, loamy type and the nature of the soil is unique for differential deposition of faeces of birds.

MAJOR FLORA AND FAUNAL DIVERSITY OF THE RAIGANJ WILDLIFE SANCTUARY AND THEIR INTERACTIONS WITH THE MIGRATORY BIRDS SPECIES

This sanctuary possesses a wide variety of plants and animals as well as bird species. Raiganj Wildlife Sanctuary with enough water bodies and trees provides a glimpse of nature with moderate wilderness.

A total of 34 species of nesting trees belonging to 21 families (Table 1) were noticed from the Raiganj Wildlife Sanctuary. On the basis of species number of the family Lythraceae, Lecythidaceae, Euphorbiaceae, Apocynaceae and Moraceae are most dominant. The mammals (2 species; 2 families), birds (29 species; 20 families) and fishes (19 species; 11 families) were recorded during the survey period (Tables- 2, 3 ,4, 5). The major plant species of the sanctuary are *Barringtonia acutangula* (Fig. 7), *Lagerstroemia speciosa* (Fig. 9), *Terminalia arjuna, Dalbergia sisoo, Streblus asper, Trewia nudiflora* (Fig. 8), *Alstonia scholaris, Bambusa tulda Ficus bengalensis, Anthocephalus indicus* (Fig. 10), *Ficus glomerata, Bombax ceiba, Swictcnia mehagoni, Sizygium cumini, Tectona grandis, Artocarpus heterophyllus, Polyalthea longifolia, Mangifera indica, Gmelina arborea* and *Caesalpinia pulcherima*. The area acquires the atmosphere of a bird sanctuary due to congregation of Open- billed stork (*Anastomus oscitans*) (Fig. 11), Heron (*Nycticorax nycticorax*) (Fig. 13), *Ardea cinera* (Fig.

19), (*Ardeola grayii*), Egret (*Bulbulcus ibis*) (Fig. 14), *Egretta garzetta, Mesophoyx intermedia* (Fig. 15) and Cormorant (*Phalacrocorax fuscicillis, Phalacrocorax niger*) (Fig. 12)) in large number. Some other common birds are Bronze-winged Jacana (*Metropidius indicus*) (Fig. 16), Common Myna (*Acridotheres tristis*) (Fig. 18) and House Crow (*Corvus splendens*) (Fig. 17). The most common amphibia were *Bufo melanostictus* (Fig. 23) and *Rana tigrina* (Table 5). About reptiles the Bengal monitor (*Varanus bengalensis*) (Fig. 24) calotes (*Calotes versicolor, Natrix piscator*) were most common (Table 5). As for mammals, Jackal (*Canis aureus)* and small Indian civet (*Viverricula indica)* can be seen sometimes coming out from outside to the sanctuary (Table 5). Among the fishes, *Esomus danricus, Amblypharyangodon mola* (Fig. 22), *Puntius punti, Puntius ticto, Lebeo rohita, Cirrhinus mrigala, Chanda nama, Trichogaster fasciatus* (Fig. 21), *Monopterus cuchia, Heteropneustes fossilis, Clarius batracus, Channa striatus, Channa gachua* were dominant fishes in the water bodies of this sanctuary (Table 3). In the present study, 5 species of molluses belonging to 5 families (Table 4) were collected and identified. *Bellamya* (Fig. 20), *Pila* and *Lamellidens* were some of the edible molluscan genera of this forest.

Raiganj Wildlife Sanctuary rich with enough waterbodies and trees and attracted many migratory birds particulary Open- billed stork, Heron, Egret and Cormorants. They came in this sanctuary mainly for their breeding purpose. The Open –billed stork (*Anastomus ocitans*) started to arrive in this sanctuary in the last week of May of the year. The arrival time of Heron, Egret and Cormorant birds was second week of the month of April.

Raiganj Wildlife Sanctuary provided a wide variety of food for the birds, particularly for the Asian open bill, whose main diet is apple snail and for Heron, Egret and Cormorant, whose main diet is fish. The Open-bills collected their maximum food from the canals of the sanctuary (Fig. 25) and paddy fields around the sanctuary. The Open- billed stork is a colonial breeder. The storks breeded with other birds (Fig. 5, 6) like Egrets, Herons and Cormorants in the sanctuary without hampering each-other (Fig. 26, 27, 28, 29, 30). With respect to nest location *Anastomus oscitans* were found nesting in both core and buffer region of the sanctuary as well as near human habitation and road side but other migratory bird species were found nesting in core region near water sources of the santary.

The tree species recorded, *Barringtonia acutangula, Lagerstroemia speciosa, Terminalia arjuna, Dalbergia sisoo, Streblus asper, Trewia nudiflora, Alstonia scholaris* and *Bambusa tulda* were found to be used as nesting tree by three bird families i.e., Ciconidae, Ardeidae and Phalacrocoracidae on the same tree (Table 8, 9) while *Ficus bengalensis, Anthocephalus indicus, Ficus glomerata, Bombax ceiba, Swietenia mehagoni, Sizygium cumini, Tectona grandis, Artocarpus heterophyllus, Polyalthea longifolia, Mangifera indica, Gmelina arborea, Caesalpinia pulcherima* and other plants were used by only *Anastomus oscitans* (Table 6, 7). The site of breeding colony of these birds in the Raiganj Wildlife Sanctuary had 645 trees. Among these plant species maximum number of nest of Open-billed stork (34.85% and 35%), Heron (77.23% and 76.88%) Egret (66.21% and 62.49%) and Cormorant (79.65% and 78.13%) were built on the tree *Lagerstroemia speciosa* of the sanctuary in the year 2007 and 2008 respectively (Table 6, 7). In all sampled trees (n = 645), in which nesting activity were observed were determined (Table 6, 7). The height of the nesting trees were ranged from 20.23 ±0.28 ft to 65.92 ± 1.29 ft. The height of the nest from the ground of these bird species varied from 10.12 ±2.38 ft to 56.81±6.29 ft. The Open - billed stork built their nest at the topmost position of the nesting tree together with other bird species of the sanctuary. About 98 % nests of Open-billed stork were situated at the junction of two or three branches and only 2 % nests were on the single thick strong branch of the trees were observed. There was minimum 1 ft distance between nearest two nests. The Open-billed storks built their nest together with Cormorants, Herons and Egrets on the same nesting tree without disturbing each other.

Sticks of *Bambusa tulda* were more used as the nesting materials by Egret, twigs of *Barringtonia acutangula, Lagerstroemia speciosa* and *Mangifera indica* were used by Heron and Cormorant, fresh leaaves and twigs of *Ficus glomerata, Eucalyptus citriodora, Lagerstroemia speciosa, Anthocephalus indicus, Cynodon dactylon, Cyperus rotundus, Croton bonplandianum* were used more commonly in formation of nest by Open - billed stork from the sanctuary. The Open - billed stork collected 29% of wet leaves and grass from canals and waterbodies, 55% twigs and fresh branches from plants and 16% of dry branches and sticks from ground of the sanctuary and neibouring areas . The peak month in nesting was observed between July and August of the year. The highest nest density of Open-billed stork was recorded on *Streblus asper* (37.00 and 25.00) and *Barringtonia acutangula* (24.00 and 25.00) in 2007 and 2008 respectively (Table 8, 9). The highest nest density of Heron, Egret and Cormorant was observed on *Bambusa tulda* in 2007 and 2008 (Table 8, 9). Open-billed stork had the maximum number of nests (8970) followed by Herons (2410), Egrets (1474) and Cormorants (1469) in the year 2007 and Open- billed stork (11784) followed by Herons (2560), Egrets (1941) and Cormorants (2021) in the year 2008 (Table 10) .

A large number of Open-billed stork (44816 and 58920), Cormorant (7348 and 10108), Heron (12052 and 12804) and Egret (7392 and 9708) had in the sanctuary in the year 2007 and 2008 respectively (Fig. 4). Shape, length, breadth and depth of the nest (Table 11) were measured and recorded in the year 2007 and 2008. The shape of the nest of Open- billed stork and Heron was semicircle, in Cormorant nest was circle and in Egret nest was oval. The female of Open-billed stork laid 2 to 5 eggs, Cormorant laid 3 to 4 eggs, Heron laid 2 to 4 eggs and Egret laid 1 to 4 eggs in one reproductive cycle (Table 12). Initially the colour of the egg was white in Open-billed stork (Fig. 31, 32) bluish in Little Cormorant (Fig. 34) and Cattle Egret (Fig -36) and bluish with brownish spot in Night Heron (Fig. 33), bluish in Little Egret (Fig. 35) and after 7 days all the eggs were transformed into brownish colour. Radius, length and weight of the eggs are presented in the table (Table 12). The shape of the eggs was oval, more or less pointed at one end and the other end was rounded. It was noticed that most of the eggs of these birds were laid at one to two days interval.

After grown up of chicks the Egret and Heron first started to leave this forest from the middle of October of the year whereas Cormorant and Open- billed stork leave this place from the middle of November of the year.

Common myna, House sparrow, Kingfisher, Red- vented bulbul, Spotted dove, Oriental magpie rabin and House crow were seen in all the month in the sanctuary in the year 2007 and 2008.

Table 1. Major plants (trees) species within the Raiganj Wildlife Sanctuary, Raiganj, India.

S.No.	Common Name	Scientific Name	Family
1.	Jarul	*Lagerstroemia speciosa* (L) Pers.	Lythraceae
2.	Hijal	*Barringtonia acutangula* (L) Gaertn.	Lecythidaceae
3.	Pituli	*Trewia nudiflora* L.	Euphorbiaceae
4.	Arjun	*Teminalia arjuna* (Roxb.) Wight & Arn.	Combretaceae
5.	Chatim	*Alstonia scholaris* L.R.Br.	Apocynaceae
6.	Seora	*Streblus asper* Lour.	Moraceae
7.	Bamboo	*Bambusa tulda* Roxb.	Poaceae

S.No.	Common Name	Scientific Name	Family
8.	Eucalyptus	*Eucalyptus citriodora* (Hook) K.D.	Myrtaceae
9.	Akashmoni	*Acacia auricuiliformis* A. Cunn & Ex Benth.	Liguminosae
10.	Kadamba	*Anthocephalus indicus* (Roxb.)	Rubiaceae
11.	Sisoo	*Dalbergia sisoo* Roxb.	Papilionaceae
12.	Dumur	*Ficus glomerata* Roxb.	Moraceae
13.	Mehagani	*Swietenia mahagoni* (L.) Jacq.	Meliaceae
14.	Bat	*Ficus bengalensis* L.	Moraceae
15.	Jamun	*Sizygium cumini* (L.) Druce	Myrtaceae
16.	Teak	*Tectona grandis* L.f.	Verbinaceae
17.	Jack fruit	*Artocarpus heterophyllus* Larn.	Moraceae
18.	Debdaru	*Polyalthea longifolia Sonn.*	Annonaceae
19.	Krishnachura	*Caesalpinia pulcherima* (L.) Sw.	Caesalpiniaceae
20.	Gamar	*Gmelina arborea* Rox	Verbinaceae
21.	Mango	*Mangifera indica* L.	Anacardiaceae
22.	Simul	*Bombax ceiba* L.	Bombacaceae
23.	Bahera	*Terminalia belerica* (Gaertn.) Roxb.	Combretaceae
24.	Shirish	*Albizia procera* Roxb.	Mimosaceae
25.	Amlaki	*Embica officinalis* Gaertn.	Euphorbiaceae
26.	Gokul	*Ailanthes grandis* Prain	Simarubaceae
27.	Bakain	*Melia azadirachta* L.	Meliaceae
28.	Neem	*Azadirachta indica* L.	Meliaceae
29.	Sidha	*Lagerstroemia parviflora* (L.) Roxb	Lythraceae
30.	Karamja	*Carissa carandas L.*	Apocynaceae
31.	Gulmohor	*Delonix regia* (Boj. Ex Hook.) Rof.	Caesalpiniaceae
32.	Water hyacinth	*Eichornia crassipes* (Mart.) Solms	Pontedriaceae

S.No.	Common Name	Scientific Name	Family
33.	Keya	*Pandamus fascicularis* Lam.	Pandanaceae
34.	Guava	*Psidium guyava* L.	Myrtaceae

Table 2. Name and number of bird species that were spoted in the Raiganj Wildlife Sanctuary, Raiganj, India

English Name	Scientific Name	Family	Number (2007)	Number (2008)
Open-billed stork	*Anastomus oscitans*	Ciconiidae	44816	58920
Night heron.	*Nycticorax nycticorax*	Ardeidae	5245	5527
Gray heron	*Ardea cinera*	Ardeidae	3709	3725
Pond heron	*Ardeola grayii*	Ardeidae	3098	3552
Little cormorant	*Phalacrocorax niger*	Phalacrocoracidae	5036	7259
Indian cormorant	*Phalacrocorax fuscicillis*	Phalacrocoracidae	2312	2849
Cattle egret	*Bulbulcus ibis*	Ardeidae	2367	3537
Little egret	*Egretta garzetta*	Ardeidae	2143	2678
Intermediate egret	*Mesophoyx intermedia*	Ardeidae	1989	2170
Large egret	*Casmerodius albus*	Ardeidae	893	1323
Common myna	*Acridotheres tristis*	Sturnidae	367	356
House crow	*Corvus spledens*	corvidae	281	235
Red –vented bulbul	*Pycnonotus cafer*	Pycnonotidae	207	210
House sparrow	*Passer domestica*	Passeridae	189	192
Black drongo	*Dicrurus macrocercus*	Dicuridae	156	140
Oriental mapie rabin	*Copsychus sanlaris*	Muscicapidae	127	117

English Name	Scientific Name	Family	Number (2007)	Number (2008)
Jungle babbler	*Turdoides striatus*	Sylvidae	119	113
Spotted dove	*streptopelia chinensis*	Columbidae	113	87
Purple sunbird	*Nectarinia asiastica*	Nectariniidae	89	70
Small blue kingfisher	*Acedo anhia*	Alcedinidae	87	93
Asian koel	*Eudynamys scolopacea*	Cuculidae	60	54
Bronze winged jacana	*Metropidius indicus*	Jacanidae	53	67
Lesser golden-backed woodpecker	*Dinpoinum benghalense*	Indicatoridae	51	50
Black headed oriole	*Oriolus xanthornus*	Oriolidae	47	34
Jungle crow	*Corvus macrorhynchos*	Corvidae	23	31
Common hoopoe	*Upopa epops*	Upipidae	21	19
Whit breasted water-hen	*Amaurornis phoenicurus*	Rallidae	19	15
Rose-ringed parakeet	*Psillacula krameri*	Pasittacidae	12	19
Asian pied starling	*sturnus contra*	Sturnidae	9	5

Table 3. Abundance of fishes recorded in the water body of the Raiganj Wildlife Sanctuary, Raiganj, India

Esomus danricus (Ham)	Cyprinidae
Amblypharyangodon mola (Ham)	Cyprinidae
Puntius punti (Ham)	Cyprinidae
Puntius ticto (Ham)	Cyprinidae
Chanda nama (Ham)	Chandidae
Chanda ranga (Ham)	Chandidae
Labeo rohita (Ham)	Cyprinidae
Cirrhinus mrigala (Ham)	Cyprinidae
Channa punctatus (Bloch)	Ophiocephalidae
Mystus tengara (Ham)	Bagridae
Channa striatus (Bloch)	Ophiocephalidae
Heteropneustes fossilis (Bloch)	Heteropneustidae
Clarius batracus (Bloch)	Claridae
Trichogaster fasciatus	Osphronemidae
Glossogobius giurus (Ham)	Gobiidae
Monopterus cuchia (Ham)	Synbranchidae
Anabus testudineus (Bloch)	Anabantidae
Channa gachua (Ham)	Ophiocephalidae

Table 4. Abundance of moulluscs that were spotted in the study spots alson with their family names.

Bellamya bengalensis	Viviparidae
Lymnaea acuminata	Lymnaeidae
Pila globasa	Pilidae
Lameliidens marginalis	Unionidae
Achatina fulica	Stenogyridae

Table 5. Amphibia, reptiles and mammals recorded in the study spots.

Amphibias	**Family**
Bufo melanosticlus	Bufonidae
Rana trigina	Ranidae
Reptiles	**Family**
Calotes versicolor	Aganidae
Varanus bengalensis	Varanidae
Naja naja	Elapidae
Natrix piscator	Colubridae
Mammals	**Family**
Canis aureus	Canidae
Viverricula indica	Viverridae

Table 6. Nesting plant species, number of nesting trees and nest numbers of the migratory birds in the colony site in the year 2007.

Plant species	Total No. of trees	Open billed storks		Herons		Egrets		cormorants	
		Trees occupied	No.of Nest (%)	Trees occupied	No.of Nest (%)	Trees occupied	No. of Nest (%)	Trees occupied	No.of Nest (%)
Ficus beng bengalensis	04	04	862 (9.61%)						
Anthocephalus indicus	13	13	231 (2.58%)						
Barringtonia acutangula	73	72	811 (9.04%)	05	20 (0.82%)	02	07 (0.47%)	04	14 (0.95%)
Ficus glomerata	04	04	65 (0.72%)						
Terminalia arjuna	55	54	434 (4.84%)	04	17 (0.69%)	01	09 (0.61%)	02	10 (0.68%)
Dalbergia sisoo	06	05	26 (0.29%)	05	36 (1.51%)	02	30 (2.04.82 %)	02	16 (1.09%)
Streblus asper	10	10	251 (2.80%)	03	23 (0.96%)	01	05 (0.34%)	01	04 (0.27%)
Lagerstroemia speciosa	284	249	3126 (34.85%)	179	1861 (77.23%)	138	976 (66.21%)	153	1170 (79.65%)

Plant species	Total No. of trees	Open billed storks		Herons		Egrets		cormorants	
		Trees occupied	No.of Nest (%)	Trees occupied	No.of Nest (%)	Trees occupied	No. of Nest (%)	Trees occupied	No.of Nest (%)
Trewia nudiflora	113	113	1804 (20.11%)	07	50 (2.06%)	05	24 (1.63%)	05	20 (1.36%)
Alstonia scholaris	34	31	269 (3%)	08	51 (2.13%)	04	19 (1.29%	07	57 (3.88%)
Bambusa tulda	15	10	35 (039%)	15	352 (14.60%)	15	404. (27.41%	14	178 (12.12%)
Bombax ceiba	04	04	91 (1.01%)						
Swietenia mahagoni	03	03	109 (1.22%)						
Sizygium cumini	02	02	46 (0.51%)						
Tectona grandis	02	02	90 (1%)						
Artocarpus heterophyllus	02	02	89 (0.99%)						
Polyalthea longifolia	01	01	18 (0.20%)						
Mangifera indica	01	01	94 (1.05%)						

Plant species	Total No. of trees	Open billed storks		Herons		Egrets		cormorants	
		Trees occupied	No.of Nest (%)	Trees occupied	No.of Nest (%)	Trees occupied	No. of Nest (%)	Trees occupied	No.of Nest (%)
Gmelina arborea	01	01	19 (0.21%)						
Caesalpinia pulcherima	01	01	40 (0.45%)						
Others Plants	17	17	460 (5.23%)						
Total	**645**	**599**	**8970**	**226**	**2410**	**168**	**1474**	**188**	**1469**

Table 7. Nesting plant species, number of nesting trees and nest numbers of the migratory birds in the colony site in the year 2008.

Plant species	Total No.of trees	Open billed stork		Herons		Egrets		Cormorants	
		Trees occupied	No.of Nest (%)	Trees occupied	No.of Nest (%)	Trees occupied	No. of Nest (%)	Trees occupied	No.of Nest (%)
Ficus bengalensis	04	03	1250 (10.61%)						
Anthocephalus indicus	13	12	246 (2.09%)						
Barringtonia acutangula	73	69	996 (8.45%)	11	38 (1.48%)	07	039 (2.01%)	10	33 (1.63%)
Ficus glomerata	04	04	83 (0.7%)						
Terminalia arjuna	55	55	130 (1.1%)	04	21 (0.82%)	03	30 (1.55%)	03	34 (1.68%)
Dalbergia sisoo	06	04	25 (0.21%)	04	25 (0.98%)	01	08 (0.41%)	02	22 (1.09%)
Streblus asper	10	10	249 (2.11%)	03	14 (0.55%)	03	20 (1.03%)	04	15 (0.74%)
Lagerstroemia speciosa	284	280	4124 (35%)	197	1968 (76.88%)	162	1213 (62.49%)	175	1579 (78.13%)

Plant species	Total No.of trees	Open billed stork		Herons		Egrets		Cormorants	
		Trees occupied	No.of Nest (%)	Trees occupied	No.of Nest (%)	Trees occupied	No. of Nest (%)	Trees occupied	No.of Nest (%)
Trewia nudiflora	113	102	2121 (18%)	12	85 (3.32%)	11	29 (1.49%)	12	37 (1.83%)
Alstonia scholaris	34	34	353 (3%)	05	26 (1.01%)	05	21 (1.08%	08	35 (1.73%)
Bambusa tulda	15	08	38 (0.32%)	15	383 (14.96%)	15	581 (29.93%	13	266 (13.16%)
Bombax ceiba	04	04	132 (1.12%)						
Swietenia mahagoni	03	03	166 (1.41%)						
Sizygium cumini	02	02	90 (0.76%)						
Tectona grandis	02	02	136 (1.15%)						
Artocarpus heterophyllus	02	02	232 (1.97%)						
Polyalthea longifolia	01	01	38 (0.32%)						

Plant species	Total No.of trees	Open billed stork		Herons		Egrets		Cormorants	
		Trees occupied	No.of Nest (%)	Trees occupied	No.of Nest (%)	Trees occupied	No. of Nest (%)	Trees occupied	No.of Nest (%)
Mangifera indica	01	01	187 (1.59%)						
Gmelina arborea	01	01	40 (0.34%)						
Caesalpinia pulcherima	01	01	63 (0.54%)						
Others Plants	17	17	1085 (9.21%)						
Total	**645**	**615**	**11784**	**251**	**2560**	**207**	**1941**	**227**	**2021**

Table 8. No. of plant species commonly occupied by major migratory birds in the year 2007.

Plant species	No. of trees ocupied by major migratory birds	No. of nest of Open Bills	No. of nest of Herons	No. of nest of Egrets	No. of nest of Cormo-rants
Lagerstroemia speciosa	138/284	1513 (10.96)*	1155 (8.37)	800 (5.80)	903 (6.54)
Barringtonia acutangula	02//73	48 (24.00)	06 (3.00)	07 (3.50)	07 (3.50)
Trewia nudiflora	05/113	24 (4.80)	35 (7.00)	14 (2.80)	15 (3.00)
Terminalia arjuna	01/55	03 (3.00)	06 (6.00)	09 (9.00)	09 (9.00)
Streblus asper	01/10	37 (37.00)	03 (3.00)	05 (5.00)	04 (4.00)
Bambusa tulda	10/15	26 (2.60)	231 (23.10)	262 (26.20)	120 (12.00)
Dalbergia sisoo	02/06	02 (1.00)	03 (1.50)	07 (3.50)	02 (1.00)

*Nest density

Table 9. No. of plant species commonly occupied by major migratory birds in the year 2008.

Plant species	No. of trees ocupied by major migratory birds	No. of nest of Open Bills	No. of nest of Herons	No. of nest of Egrets	No. of nest of Cormo-rants
Lagerstroemia speciosa	162/284	2430 (15.00)	1428 (8.81)	972 (6.00)	1190 (7.34)
Barringtonia acutangula	07//73	175 (25.00)	28 (4.00)	35 (5.00)	28 (4.00)
Trewia nudiflora	11/113	80 (7.27)	72 (6.54)	22 (2.00)	35 (3.18)

Plant species	No. of trees ocupied by major migratory birds	No. of nest of Open Bills	No. of nest of Herons	No. of nest of Egrets	No. of nest of Cormo-rants
Terminalia arjuna	03/55	07 (2.33)	15 (5.00)	24 (8.00)	30 (10.00)
Streblus asper	03/10	75 (25.00)	10 (3.33)	18 (6.00)	10 (3.33)
Bambusa tulda	11/15	30 (2.73)	304 (27.64)	410 (37.27)	168 (15.27)
Dalbergia sisoo	01/06	03 (3.00)	04 (4.00)	06 (6.00)	05 (5.00)

*Nest density

Table 10. Total No.of Nest of major migratory birds recorded in the year 2007 and 2008 in the Raiganj Wildlife Sanctuary.

Year	Open Bills	Cormorants	Herons	Egrets
2007	8970	1469	2410	1474
2008	11784	2021	2560	1941

Table 11. Description of nest of major migratory birds in the Raiganj Wildlife Sanctuary

Variables	No. of nest studied	*Anastomus oscitans*	*Phalacro-corax niger*	*Nycticorax nycticorax*	*Bulbulcus ibis*
Shape	100	Semicircle	Circle	Semicircle	Ovale
Length (cm)	100	50.30±0.11	33.45±0.28	36.05±0.07	31.15±0.21
Breadth (cm)	100	40.90±0.07	29.1±0.09	32.54±0.03	23.4±0.36
Depth (cm)	100	6.90±0.02	4.48±0.34	6.08±0.57	5.02±0.04

Table 12. Description of eggs of major migratory birds in the Raiganj Wildlife Sanctuary.

Variab-les	**No. of nest studied**	***Anastomus oscitans***	***Phalacro-corax niger***	***Nycticorax nycticorax***	***Bulbulcus ibis***
Number of Eggs/ Nest	100	2-5	3-4	2-4	1-4
Colour	100	White	Bluish	Bluish with brownish spot	Bluish
Weight (g)	100	50.56±0.01	29.24±0.89	25.11±0.13	19.71±0.10
Length (cm)	100	06.07±0.004	5.6±0.23	5.21±0.007	4.74±0.09
Radius (cm)	100	03.10±0.002	3.05±0.05	2.66±0.008	2.28±0.70

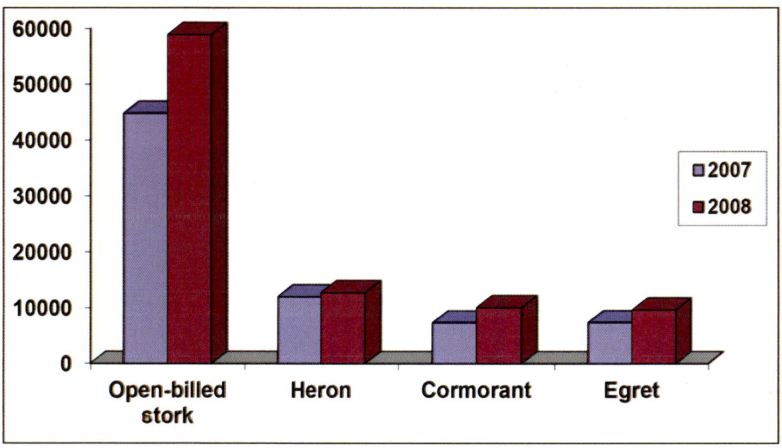

Fig. 4: Number of most dominant major migratory birds recorded during the study period (2007-2008) in the Raiganj Wildlife Sanctuary.

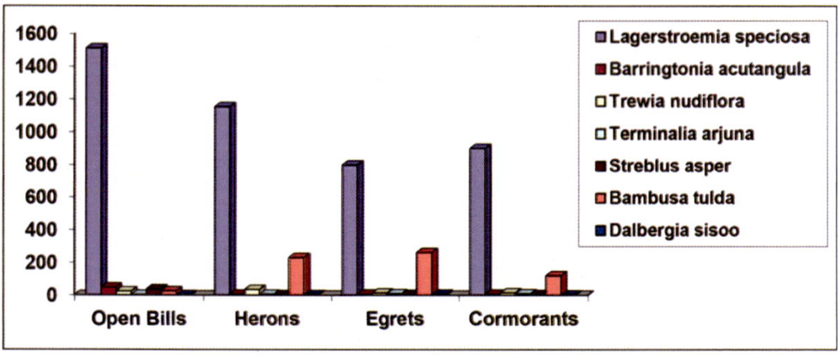

Fig. 5. Association of attributes of Open-bills with other migretory birds according to the number of nest / nesting tree species in the sanctuary in the Year 2007.

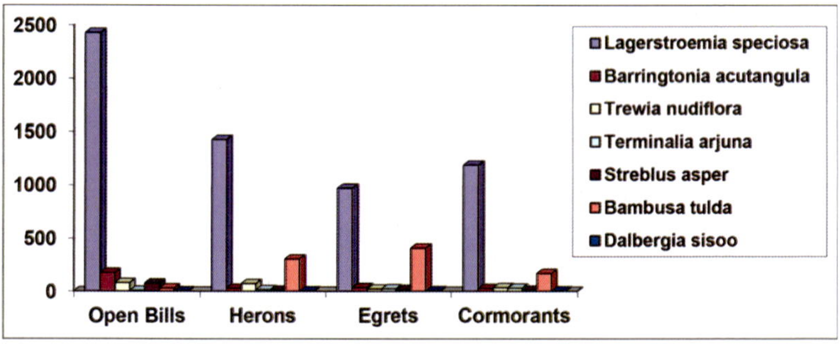

Fig. 6. Association of attributes of Open-bills with other migretory birds according to the number of nest / nesting tree species in the sanctuary in the Year 2008.

Fig. 7. *Barringtonia acutangula*

Fig. 8. *Trewia nudiflora*

Fig. 9. *Lagerstroemia speciosa*

Fig. 10. *Anthocephalus indicus*

Fig. 11. Anastomus oscitans

Fig. 12. *Phalacrocorax niger*

Fig. 13. *Nycticorax nycticorax*

Fig. 14. *Bulbulcus ibis*

Fig. 15. *Mesophoyx intermedia*

Fig. 16. *Metropidius indicus*

Fig. 17. *Corvus spledens*

Fig. 18. *Acridotheres tristis*

Fig. 19. *Ardea cinera*

Fig. 20. *Bellamya bengalensis*

Fig. 21. *Trichogaster fasciatus*

Fig. 22. *Amblypharyangodon mola*

Fig. 23. *Bufo melanosticlus*

Fig. 24. *Varanus bengalensis*

Fig. 25. Open-billed storks collected food (molluscs) from the canal of the Raiganj Wildlife Sanctuary.

Fig. 26. Open-billed storks took rest together with Cormorant and Cattle Egret on *Barringtonia acutangula*.

Fig. 27. Open-billed storks built nest together with Herons and Egrets on *Lagerstroemia speciosa.*

Fig. 28. Open-billed storks built nest together with Cormorants and Egrets on *Bambusa tulda.*

Fig. 29. Open-billed storks built nest together with Cormorants on *Lagerstroemia speciosa.*

Fig. 30. Open-billed storks built nest together with Herons and Cormorants on *Lagerstroemia speciosa.*

Fig. 31. The eggs on the nest of Open-billed stork.

Fig. 32. The eggs with chicks on the nest of Open-billed stork.

Fig. 33. The eggs on the nest of Night Heron.

Fig. 34. The eggs on the nest of Cormorant.

Fig. 35. The eggs with chick on the nest of Little Egret

Fig. 36. The eggs on the nest of Cattle Egret.

Species composition of birds in an area is related to the type of vegetation, height above the sea level, availability of microhabitats and various other factors. The migratory birds were coming to this sanctuary mainly for nesting and breeding purpose. Open-billed stork, Heron, Egret and Cormorant utilized the water bodies to feed on fish and mollusca. Open-billed stork mainly built their nests on the tree *Lagerstroemia speciosa,* because this trees are branched, hard and bushy type, helps to construct the nest properly. The characteristics of sites where a bird species selects to nest are important since the birds depend on both protections from predators. A preliminary analysis of host- plant characteristics showed that tree species with large, dense canopies were preferred by birds for placing their nests. A dense canopy cover reduces the thermal stress to vulnerable young and provides hide from the potential air, predators.

Open- billed stork built their nests together with other birds like Egrets, Herons and Cormorants in the core region of the sanctuary as though the stork is a colonial breeder. Egrets preferred mainly *Bambusa tulda* tree, Herons and Cormorants built their nest mainly on *Barringtonia acutangula, Lagerstroemia speciosa, Bambusa tulda* tree. It was noticed that the population of Egrets, Herons and Cormorants were more in abundance in the core region than the buffer region but the Stork population were seen both in the core region as well as buffer region as they can tolerate human interference.

Nesting association of open-billed stork with Herons, Egrets and Cormorants may be due to shortage of nesting sites. Such an association is reported in Indian Black Ibis (*Pseudibis papillosa*) inhabiting the arid zone of Rajsthan and is also observed in other ibises. The nesting trees selected for nesting near the water resources were seen in Open- billed stork as well as in Heron, Egret and Cormorant. This type of nesting was observed in little Cormorant (*phalacrocorax niger*). The nesting trees near the water resources were fully occupied by early comer Storks, Herons. Egrets and Cormorants come in the sanctuary within mid of June. So the late comer Open- billed storks preferred to nest on the trees along with the road (National Highway – 34) which runs southern border of the sanctuary inspite of their exposure to wind and noise of traffic. It may be availability of preferred nesting trees, nesting materials and feeding grounds. This type of nesting was observed in Indian Black Ibis in Rajasthan.

Sticks of *Bambusa tulda* were more used as the nesting materials by Egret and twig of *Barringtonia acutangula Lagerstroemia speciosa* and *Mangifera indica* were preferred for nest construction by Heron and Cormorant. Fresh leaves of Dumar (*Ficus glomerata*), Eucalyptus (*Eucalyptus citriodora*) Jarul (*Lagerstroemia speciosa*), Kadamba (*Anthocephalus indicus*), Durba grass (*Cynodon dactylon*) and Mutha grass (*Cyperus rotundus*) were used more commonly in formation of nest by Open -billed stork. The nest materials of Open-bills consisted of dry or wet leaves, green branches and sticks and dry or wet grasses. The Open-billed stork collected this preferred nesting materials mostly from the canals, plants and ground of the sanctuary in the same breeding season with other birds like Herons, Egrets and Corononorants without disturbing each other. This study has also revealed that there was a competition between Ciconiidae, Ardeidae and Phalacrocoracidae species regarding tree species and nesting materials. The birds used a variety of nesting materials for nest construction and most preferred soft fibers, grasses, twigs, sticks and leaves .

Both of the partners of Open- billed stork as well as Heron, Egret and Cormorant took part of the collection of the nest materials and the arrangement of the materials in the nest. The ovserverd participation of both sexes of Little Cormorant in nest building was mentioned by many scientists.

The shape of the nest of Open-billed stork and Night Heron was semicircle, in Little Cormorant nest was circle and in Cattle Egret was oval. The female of Open-billed stork laid

2 to 5 eggs and initially the colour of the egg was white. The colour of the egg was bluish in Little Cormorant and Cattle Egret and bluish with brownish spot in Night Heron. Generally white eggs are formed by calcium carbonate; the pigments biliverdin and its zinc chelate give a blue or green and protoporphyrin produces reds and browns as the ground colour. For example the hole nesting species tended to have immaculate white eggs, ground nesters surrounded by vegetation laid eggs of obscured brown, gray or olive.

The nest density of Open-billed stork, Heron, Egret and Cormorant per tree in the sanctuary is large on *Streblus asper, Bambusa tulda, Barringtonia acutangula* and *Lagerstroermia speciosa*. The nesting densities differed depending on the preferred tree species of the birds. Nesting was initiated by the Ardeid species and then Phalacrocoracidae species who had the advantage of choosing from all the potential nest sites and the opportunity to occupy the most suitable ones. Though Open-billed stork came late in the sanctuary, their larger body size helped to get nesting site within the Ardeid and Phalacrocoracidae species. The Storks always built nests at the top level of trees. This type of nesting behaviour of Storks minimized the nesting competition. House crow (*Corvus splendens*) and Bengal monitor (*Varanus bengalensis*) were the main predator of the eggs and nestlings of the Open-billed stork and other birds. It is surprising that the population of Open-billed stork, Heron, Cormorant and Egrets increased than other resident birds. It may be due to availability of nesting trees and food preferred by these particular species in the sanctuary. Raiganj Wildlife Sanctuary is a flood plain area and main rice growing area of the region dotted with many large water bodies, river beds and marshy areas which are full of main food diet, *Pila globosa* of the Open-billed stork and fishes of Heron, Egret and Cormorant. Nest construction and placement are depended with the breeding season, suitable nest sites, nesting materials availability, food availability and predator's interaction. It can be said that nesting competition between the species can be tolerated by the abundance of food sources. Sustainable utilization of molluscs and fishes are also essential because these serve as food for many birds inhabiting the sanctuary area. Hence, flexibility has been developed to avoid any potential competition for the breeding as well on the restricted foraging ground. This could be compared with the mixed colonial Egrets and Herons, which also maintain a certain distance among their nests depending upon their body size for the comfort and to avoid competition.

Species composition of birds in an area is related to the type of vegetation, height above the sea level, availability of microhabitats and various other factors. Vegetation structure is an important factor for the choice of a specific nesting site among Ardeids. The nesting site selection is an important component of breeding behaviour. The availability of suitable feeding resources in the surrounding area of Raiganj Wildlife Sanctuary and safe nesting sites attracted these birds to this sanctuary every year.

FACTORS AFFECTING THE POPULATION OF ASIAN OPEN-BILLED STORK, THE MAJOR MIGRATORY BIRD IN THIS SANCTUARY

The Open-billed stork used to build their nests on 599 trees in 2007 and on 615 trees in 2008 in the Raiganj Wildlife Sanctuary. The total number of nests of Open-billed stork were 8970 in 2007 and which increased as11784 in 2008. The maximum number of nests of stork (34.85% in 2007 and 35.00% in 2008) were built on the tree *Lagerstroemia spaciosa* in this sanctuary. The nest density of stork was recorded 14.97 in 2007 and 19.16 in 2008. A total of 44816 storks were counted in 2007 which raised as 58920 storks in 2008 (Table 13).

Breeding occurred in all the nest that were constructed in the core region and in the buffer region of the sanctuary, but the average number of fledged young on nests at different locations was different in 2007 (Table 14) and in 2008 (Table 15). In two breeding season, the fledgling success was greater at the nests that were located in the core region (2.96 in 2007 and 3.07 in 2008) than in the buffer region (2.43 in 2007 and 2.77 in 2008) of the sanctuary.

Clutch size varied from 2 to 5 eggs but most clutches contained 3 to 4 eggs. In thc year 2007, the percentage of hatching success was 95,92,93 and 89.2 and the percentage of fledging success was 94.74, 91.30, 81.72 and 76.68 for 2,3,4 and 5 clutch sizes respectively (Table 16). The percentage of hatching success was recorded as 98,96,95, and 92.8 and the percentage of fledging success was 95.92, 93.75, 85.79 and 82.76 for 2,3,4, and 5 clutch sizes in 2008 respectively (Table 17). For both year, the percentage of hatching success and fledging success were more in small clutches that in large clutches.

The reasons for different hatching success were mainly unhatching (Fig. 40), predation (Fig. 41), strom and heavy rain. The percentage of unhatched eggs (4% in 2007 and 3% in 2008) was more than other causes of egg loss in both year 2007 and 2008 (Fig. 37). The causes of mortality of nestlings were bad weather (strom and heavy rain) (Fig. 43, 44) predation (Fig. 42) and fall down during first flight (Fig. 45). The percentage of nestlings death caused by predation, bad weather (strom and heavy rain) and fall down during first flight were 3.27, 8.09 and 4.67 in 2007 and which decreased as 2.26, 6.32 and 3.46 in 2008 respectively (Fig. 38). The legs and wings of chick were broken by falling down from the nest during first flight, storm and heavy rain. The main natural predators of the species in the sanctuary were House crow (*Corvus splendens*) (Fig. 42) and Bengal monitor (*Varanus bengalensis*). House crows were opportunistic in plundering the eggs from the nest or carry away the newly hatched chicks when both the parents were away from the nest.

The breeding behaviour of Open-billed strok was affected by entrance of tourists and villagers in the sanctuary. Tourists were disturbing by throwing stones, by shouting and by taking photographs from the breeding nest of stork. The minimum and maximum average number of tourists per day entered into the buffer region of the sanctuary were 52.15 ± 0.350 and 135.35 ± 0.472 in the month of May and December of the year respectively (Table 18). Not a single tourist was observed in the core region of the sanctuary during breeding season. But the tourists enjoyed by doing picnic in close areas of the sanctuary during breeding season of the storks that hampered their breeding life (Fig. 48). The villagers mainly came to collect fire wood from the sanctuary (Fig. 46). They invaded in the core region in the month of May (02.85 ± 0.131 villagers/month), June (02.65 ± 0.167 villagers/month), July (01.75 ± 0.160 villagers/month), November (02.45 ± 0.135 villagers/month) and December (04.25 ± 0.143 villagers/month) of the year. The minimum and maximum average number of villagers per month entered into the buffer region were 03.05 ± 0.170 and 08.15 ± 0.209 in the month of July and December of the year (Table 19).

During breeding season, a huge number of vehicles like bus, truck, private car, motor bike, van and cycles were passed on the Notional Highway-34 which is situated in the heart of the sanctuary (Fig. 47). Among recorded vehicles, the highest percentage of truck (37.3) moved on the road in the morning but the highest percentage of bus (21.03) were recorded in the evening of the day (Fig. 39).

Table 13. Total number of nesting trees occupied by Open-billed storks, total number of nests, nest density and total number of Open-billed stork population recorded in the Raiganj Wildlife Sanctuary in the year 2007 and 2008.

Year	Total number of nesting trees	Total number of nests	Nest density	Total number of Open-billed storks
2007	599	8970	14.97	44816
2008	615	11784	19.16	58920

Table 14. Breeding values of nests on different nesting places in the year 2007

Nesting places	Number of nest occupie d (A)	Number of nests with breeding (B)	% of nests with breeding (B/A)	Number of youngs fledged (C)	Average number of fledged young in different nesting places (C/B)
Core region	100	100	100	296	2.96
Buffer region	100	100	100	243	2.43

Table 15. Breeding values of nests on different nesting places in the year 2008

Nesting places	Number of nest occupied (A)	Number of nests with breeding (B)	% of nests with breeding (B/A)	Number of youngs fledged (C)	Average number of fledged young in different nesting places (C/B)
Core region	100	100	100	307	3.07
Buffer region	100	100	100	277	2.77

Table 16. Breeding success according to different clutch size of Open-billed stork in the year 2007

Clutch size	Number of observation		Number of eggs hatched	Number of youngs fledged	% of hatching success	% of fledging success
	Nests	Eggs				
2	50	100	95	90	95.00	94.74
3	50	150	138	126	92.00	91.30
4	50	200	186	152	93.00	81.72

Clutch size	Number of observation		Number of eggs hatched	Number of youngs fledged	% of hatching success	% of fledging success
	Nests	Eggs				
5	50	250	223	171	89.20	76.68
Total	200	700	642	539	91.71	83.96

Table 17. Breeding success according to different clutch size of Open-billed stork in the year 2008.

Clutch size	Number of observation		Number of eggs hatched	Number of youngs fledged	% of hatching success	% of fledging success
	Nests	Eggs				
2	50	100	98	94	98.00	95.92
3	50	150	144	135	96.00	93.75
4	50	200	190	163	95.00	85.79
5	50	250	232	192	92.80	82.76
Total	200	700	664	584	94.86	87.95

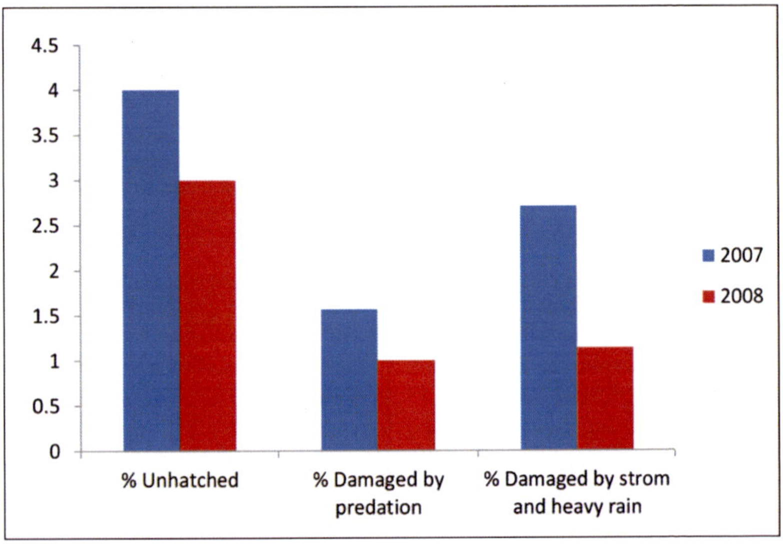

Fig. 37. Percentage of unhatched eggs and percentage of damaged eggs of Open-billed stork by different factors in the Raiganj Wildlife Sanctuary in 2007 and 2008.

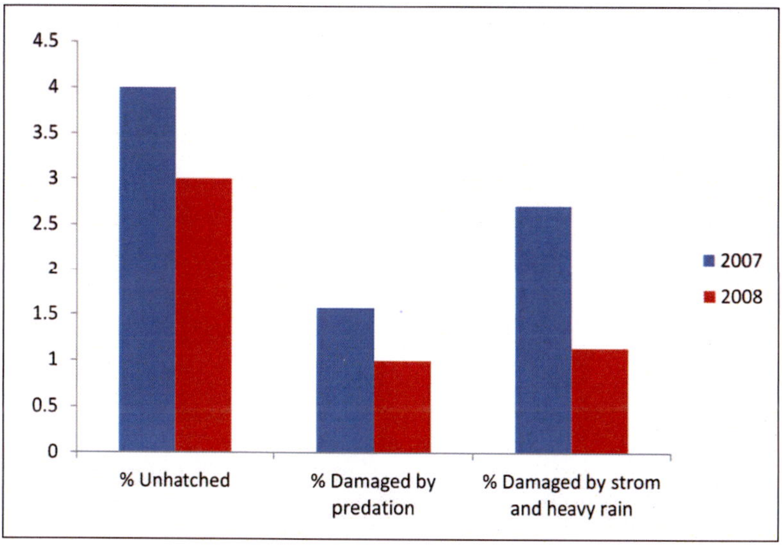

Fig. 38. Percentage of nestlings death of Open-billed stork by different factors in the Raiganj Wildlife Sanctuary in 2007 and 2008.

Table 18. Average number of tourists per day entered into the Raiganj Wildlife Sanctuary at different month of the year during breeding season of Open-billed stork in two years.

Months	Number of observation (days)	Average number of tourists / day	
		Buffer region	Core region
May	20	52.15 ± 0.350	00.00 ± 0.000
June	20	53.35 ± 0.499	00.00 ± 0.000
July	20	55.85 ± 0.327	00.00 ± 0.000
August	20	66.75 ± 0.347	00.00 ± 0.000
September	20	86.45 + 0.426	00.00 ± 0.000
October	20	105.65 ± 0.499	00.00 ± 0.000
November	20	116.85 ± 0.406	00.00 ± 0.000
December	20	135.35 ± 0.472	00.00 ± 0.000

Table 19. Average number of villagers per month entered into the Raiganj Wildlife Sanctuary at different month of the year during breeding season of Open-billed stork in two years .

Months	Number of observation (days)	Average number of villagers / month	
		Buffer region	Core region
May	20	05.85 ± 0.182	02.85 ± 0.131
June	20	04.55 ± 0.211	02.65 ± 0.167
July	20	03.05 ± 0.170	01.75 ± 0.160
August	20	04.25 ± 0.143	00.00 ± 0.000
September	20	05.15 ± 0.182	00.00± 0.000
October	20	05.95 ± 0.198	00.00± 0.000
November	20	06.75 ± 0.216	02.45 ± 0.135
December	20	08.15 ± 0.209	04.25 ± 0.143

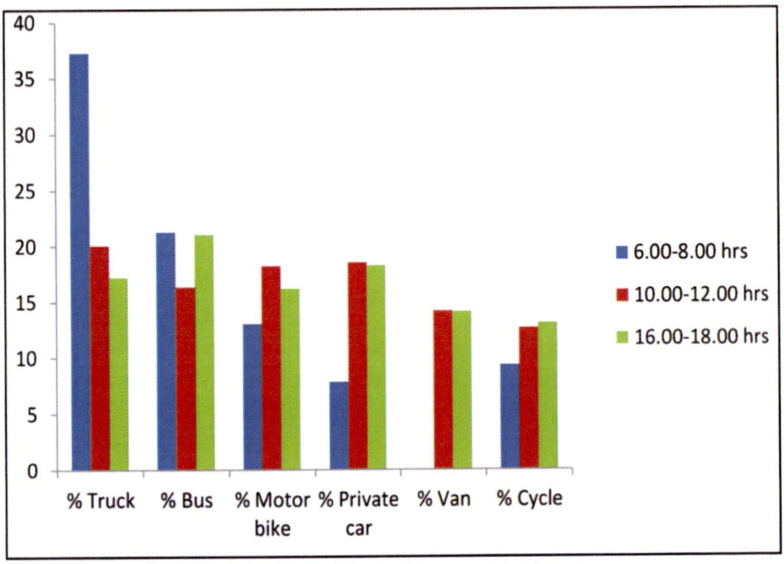

Fig. 39. Percentage of different vehicles passed on National Highway-34 through the Raiganj Wildlife Sanctuary at different time of the day during breeding season of Open-billed stork.

Fig. 40. The unhatched egg of Open-billed stork dropped on the ground of the sanctuary.

Fig. 41. The egg of Open-billed stork was damaged by predation in the sanctuary.

Fig. 42. A crow came on the nest of Open-billed stork for predation in the sanctuary.

Fig. 43. The chick of Open-billed stork was died in the sanctuary during strom and heavy rain.

Fig. 44. The chick of Open-billed stork was fallen down on the ground of the sanctuary during strom and heavy rain and the legs of the chick were broken.

Fig. 45. The chick of Open-billed stork was fallen down on the ground of the sanctuary during first flight and the chick was without parental care.

Fig .46. The villager (women) was collecting fire wood from the sanctuary during breeding season of Open-billed stork .

Fig. 47. The trucks were passing during breeding season of Open-billed stork on National Hihgway–34 which was situated in the middle of the sanctuary.

Fig. 48. Tourists were doing picnic in the buffer region of the sanctuary in the month of December of the year.

CONCLUSION

Raiganj Wildlife Sanctuary is rich with enough waterbodies, flora and fauna. A total of 34 species of trees belonging to 21 families contribute to the main flora of the sanctuary. The broad-leaved deciduous trees provide the nesting place of many migratory and resident birds species. The main migratory birds of the sanctuary are Open-billed storks, Herons, Egrets and Cormorants. Apart from the avifauna (29 species, 20 families) a large number of molluscs (5 species, 5 families), fishes (19 species, 11 families), amphibians (2 species, 2 families), reptiles (4 species, 4 families) and mammals (2 species, 2 families) are common in this sanctuary.

The Open-billed storks breed with other birds like Egrets, Herons and Cormorants in the sanctuary as though the stork is a colonial breeder. They used to come in this sanctuary mainly for their breeding purpose. The Open-billed storks started to arrive in this sanctuary from the last week of May of the year. The arrival time of Egret, Heron and Cormorant birds was second week of the month of April of the year. With respect to nest location, *Anastomus oscitans* were found nesting in both core and buffer region of the sanctuary as well as near human habitation and roadside but other migratory bird species were found nesting in the core region near water sources of the sanctuary. More than 645 trees were present within the breeding colony of these birds. The Open-billed stork mainly built their nests on the tree *Lagerstroemia speciosa, Barringtonia acutangula, Ficus glomerate, Anothocephalus indicus* etc. and Egrets preferred mainly *Bambusa tulda*, Herons and Cormorants built their nests on *Barringtonia acutangula, Lagerstroemia speciosa, Bambusa tulda* etc. Sticks of *Bambusa tulda* were more used as the nesting materials by Egret, twigs of *Barringtonia acutangula, Lagerstroemia speciosa* and *Mangifera indica* were used by Heron and Cormorant, fresh leaves and twigs of *Ficus glomerata, Eucalyptus citriodora, Lagerstroemia speciosa, Anothocephalus indicus, Croton bonplandianum, Cynodon dactylon* were used more commonly in the formation of nest by Open-billed stork. The higest nest density of Open -billed stork was on *Streblus asper* and *Barringtonia acutangula* in the year 2007–2008. The nest density of Heron, Egret and Cormorant per tree in the santuary is large on *Bambusa tulda*. The shape of the nest of Open-billed stork and Heron was semicircular, in Cormorant nest was circular and in Egret nest was oval. The shape of the eggs of these birds was oval, more or less pointed at one end and the other end was rounded. The most of the eggs of these brids were laid at one to two days interval. House crow (*Corvus splendens*) and Bengal monitor (*Varanus bengalensis*) were the main predator of the eggs and nestlings of the open-billed stork and other birds. Open-billed stork utilized waterbodies to feed on molluscs mainly apple snail and other birds on fish from the santuary and nearer water bodies.

A large number of Open-billed stork (44816 and 58920), Cormorant (7348 and 10108), Heron (12052 and 12804) and Egret (7392 and 9708) were in the sanctuary in the year 2007 and 2008 respectively. After grown up of chicks, the Egret and Heron first started to leave this forest from the middle of October of the year where as Cormorant and Open-billed stork leaved this place from the middle of November of the year.

The Open-billed storks were involved for breeding with other birds simultaneously on the same nesting tree of the sanctuary without disturbing each other and they used nesting materials and food items in the same breeding season from the sanctuary and neighbouring areas without hampering each other. The enriched diversity of flora and fauna of the sanctuary provided suitable nesting trees, nesting materials and food for Open-billed storks and other birds and attracted them to come in this sanctuary every year for their successful breeding purposes.

In 2007, 599 trees were selected by the storks for nesting of which 249 trees were *Lagerstroemia spaciosa* and in 2008, 615 trees were selected for nesting of which 280 trees were *Lagerstroemia spaciosa*. It indicates that more number of Open-billed stroks were aggregated in the sanctuary in the year 2008 than in the year 2007. The birds nesting in heronries prefer trees with dense canopy cover. The *Lagerstroemia spaciosa* tree species was most preferable nesting tree of storks because the trees are branched, busy type and hard, helps to construct the nest properly. The total number of nests and nest density were 8970 and 14.79 in the year 2007 and which increased as 11784 and 19.16 in the year 2008 respectively.

The main food species of Open-billed stork's chicks was *Bellamya bengalensis* and the major food species of *P. globosa*. The foraging areas of this bird species were a flood plain area and rice growing area of the region dotted with many large water bodies like ponds, river beds and marshy areas which were full of their main food diet *Pila globosa* and *Bellamya bengalensis*. The availability of these food species may be the cause of rising population of the Open-billed stork in the Raiganj Wildlife Sanctuary.

The average number of fledged young on nests at different locations of the sanctuary was different in the year 2007 and 2008. In two breeding seasons, the fledgling success was greater at the nests that were located in the core region (2.96 in 2007 and 3.07 in 2008) than in the buffer region (2.43 in 2007 and 2.77 in 2008) of the sanctuary. It can be suggested that core region was more preferred by the Open-billed stork than buffer region of the sanctuary due to minimum disturbance occurred by the tourists and villagers and surrounded by water canals which were rich in food of this bird species. Breeding birds require safe nesting site as well as sufficient quantity and quality of foraging habitat.

In the present study, the clutch size varied from 2 to 5 eggs and most of the nests contained 3 to 4 eggs. This variation of clutch size may be due to female's reproductive capacity, climatic condition and food supplies during breeding period. A decrease in food supplies and variation in climatic conditions result in a decrease in breeding performance. Females inherit the ability to vary clutch size within a certain range and that the upper limit of clutch size is firmly fixed. The seasonal variation in food supplies during the breeding period in the colony site was also thought to be the reason for clutch and egg size differences between breeding attempts in House Sparrow (*Passer domesticus*). Clutch size influenced the hatching succeed and fledging success of Open-billed stork in the Raiganj Wildlife Sanctuary. The percentage of hatching success and fledging success were more in small clutches that in large clutches in both year 2007 and 2008. The overall hatching success of Open-billed stork was 91.71% and 94.86% in the year 2007 and 2008 respectively. The overall fledging success of Open-billed stork was 83.96% and 87.95% in the year 2007 and 2008 respectively.

In the present study, the hatching success depended on the number of unhatched eggs, eggs lossed by predation, storm and heavy rain. The percentage of unhatched eggs, damaged eggs by predation, strom and heavy rain was higher in the year 2007 than in the year 2008. The nestlings death mainly caused by strom and heavy rain (8.09 % in 2007 and 6.32% in 2008) and by dropping during first flight (4.67% in 2007 and 3.46% in 2008). The nestlings also lossed by predation (3.27% in 2007 and 2.26% in 2008). House crows were opportunistic in plundering the eggs from the nest or carry away the newly hatched chicks when both the parents were away caused by human activities in the sanctuary. The chicks that survived the accidental falls during first flight from the nest were taken up for food by domestic cats and dogs in the sanctuary. As the nesting colony of the buffer region was localed in front of office building, forest bunglow, tourist tower and tourist lodge with human activities throughout the day, dogs and cats regularly visited this area.

During heavy rain, many nestlings fell down and were easy prey to the ground predators. Large colonies of nesting of Open-billed storks had recently become an important tourist attraction in the Raiganj Wildlife Sanctuary. The minimum and maximum average numbers of tourists per day entered into the buffer region of the sanctuary were 52.15 $\pm$ 0.350 and 135.35 $\pm$ 0.472 in the month of May and December of the year respectively. The villagers also invaded to collect fire wood into the buffer region as well as core region of the sanctuary during breeding season. It indicated that visits to this sanctuary were uncontrolled and there was growing concern about disturbance to the breeding storks.

Beside human disturbances, a huge number of vehicles like truck, bus, private car, motor bike, van and cycle were passed through the heart of the sanctuary during breeding season as the National Highway-34 runs through the sanctuary. Nest failure may be occurred by human interferences and noise pollution of vehicles. Surprisingly the Open-billed storks breeded with an increasing population in the year 2008 (58920 storks) than in the year 2007 (44816 storks) because this species can tolerate human interference and and horn of vehicles.

It is clear that there were many disturbances caused by human activities and vehicles during breeding season, though they can to breed in this sanctuary because they can tolerate these types of disturbances.

However, it is essential to take immediate action for habitat improvement and minimize disturbances to increase the population of this particular bird species.

The following steps are required for better management.

1) Precautionary measures included digging of at least two deep canals arround the Sanctuary to stop poaching of birds.
2) Some of the bird population lives outside of the protected areas, amongs local communities in the surrounding farm-lands. So, the areas of the Sanctuary should be increased.
3) Advanced training by the experts for staff of the Sanctuary should be developed every year.
4) Bird conservation programmes should be arranged to raise the bird population of the Sanctuary.
5) A new patrol based monitoring system should be implemented to give protection from poachers.
6) Consciousness about the growing bird population of the Sanctuary among the surrounding villagers should be increased.
7) Expert Veterinarians should be appointed for bird care and breeding.
8) NGOs should be engaged to educate the local people to save these migratory birds.
9) SMART (Spatial Monitoring And Reporting Tool) should be started to monitor bird populations as well as their disease.
10) Stern action by law and order should be implemented against the poachers and outsiders.

REFERENCES

Ali, S. (1977). The Book of Indian Birds (10th Edition). Bombay Natural History Society, Bombay, India.

Ali, S. and Futehally, L. (1967). Common Birds. National Book Trust, New Delhi, India.

Ali, S. and Ripley, S. D. (1986). Handbook of the birds of India and Pakistan, vol. V, Oxford University press, Bombay.

Antwi, L. A. K. and Ofori-Danson, P. K. (1993). Limnology of a tropical reservoir (The Kpong Reservoir in Ghana). *Trop. Ecol.*, 34(1): 75 -87.

APHA, (1995). Standard methods for the examination of waste water. 14th edition, APHA, AWIWA-WPCHCF, Washington DC.

Arendt, W. J. and Arendt, A. I. (1988). Aspects of the breeding biology of the Cattle Egret (*Bubulcus ibis*) in Montserrat, West Indies, and its impact on nest vegetation. Colonial Water Birds. 11 (1): 72-84.

Aslan, A. and Yavuz, M. (2010). Clutch and egg size variation, and productivity of the House Sparrow (*Passer domesticus)* : effects of temperature, rainfall, and humidity. Turk J Zool. 34:255-266.

Ayas, Z. (2008). Nest site characteristics and nest densities of Ardieds (Night Heron: *Nycticorax nycticorax,* Grey Heron: *Ardea cinerea,* and Little Egret: *Egretta garzetta*) in the Nallihan Bird Sanctuary (Sariyar Reservoir, Ankara, Turkey). Turk. J. Zool. 32: 167-174.

Baxer, G. S. (1994). The influence of synchronous breeding natal tree position and rainfall on egret nesting success. Colonial Waterbirds. 17: 120-129.

Bhatnagar, S. P., Shukla, S. K. and Bhaum, M. S. 2(004). Studies on the distribution of the heronries in and around Ajmer city ecosystem. Newsletter for Birdwatchers 44(3) : 46-47.

Bolger, D. T., Patten, M. A. and Bostock, D. C. (2005). Avian reproductive failure in response to an extreme climatic event. Oecologia, 142: 398-406.

Boulenger G. A. (1890). *Fauna of British India, Including Ceylon and Burma. Reptilia and Batrachia.* London.

Bouton, S. N. , Frederick, P. C., Rocha, C. D., Santos, A.T. B. D. and Bouton, T.C. (2005). Effects of tourist disturbance on Wood Stork nesting suc-cess and breeding behaviour in the Brazilian Pantanal. Waterbirds 28(4): 487-497.

Boyd, W. S., Savard, J. P. L. and Smith, G. E. J. (1989). Relationship between aquatic Birds and wetland characteristics in the aspen park-lands, Central British Columbia. Lake and Reservoir Mgmt. 5: 133 - 141.

Burger, J. (1978). The pattern and mechanism of nesting in mixed species heronries In: Wading Birds (eds. A. Sprunt, J. C. Odgen and S. Wincker). National Audubon Society Research Report No. 7. New York. Pp. 45-58.

Burger, J. (1982). On the nesting location of Cattle egret Bubulcus ibis in South African heronries. Ibis. 124: 523-529.

Carney, K. M. and Sydeman, W. J. (1999). A review of human disturbance effects on nesting colonial waterbirds. Waterbirds, 22: 68-79.

Chinery, M. (1966). *A Pictorial Dictionary of the Animal world.* London: Sampson Law, Marston and Company Ltd.

Datta, T. (1992). Behavioural ecology of feeding, reproduction in open – billed stork (*Anastomus oscitans*) (Boddaert) at the Raiganj wildlife Sanctuary, Raiganj, West Dinagpur, India. North Bengal University, India (Ph. D. thesis).

Datta, T. and Pal, B. C. (1993). The effect of human interference on the nesting of the open bill stork *Anastomus oscitans* at the raiganj wildlife sanctuary, India. Biological Conservation 64 (2) : 149 – 154.

Detenbeck, N. E., Taylor, D. L., Lima, A. and Hagley, C. (1996). Temporal and spatial variability in water quality of wetlands in the Minneapolis / St. Paul, MN metropolitan area: Implications for monitoring strategies and designs. *Environ. Monit. Assess.*, 40(1): 11 - 40.

Dobrowolski, K. A., Halba, R. and Nowicki, J. (1976). The role of birds in eutrophication by import and export of trophic substances of various waters. *Limnologica.*, 10: 543 - 549.

DuBowy, P. J. (1988). Waterfowl communities and seasonal environments: temporal variability in interspecific competition. Ecology. 69: 1439 - 1453.

Fasola, M. and Alieri, R. (1992). Nest site characteristics in relation to body size in herons in Italy. Colonial Waterbirds. 15: 185-192.

Fleming, R. L. Sr., Fleming. R. L. Jr., Bangdel. L. S. (2000). *Birds of Nepal with Reference to Kashmir and Sikkim*. Delhi: First Adarsh Impression, Gaurav Offset. Francis, London.

Grierson, A. J. C. , Long, D. G. (1983). *Flora of Bhutan*. Volume 6. Edinburgh: Royal Botanical Garden

Gunther, A. (1864). *The reptiles of British India*. Bombay: Oxford and IBH Publishing Co.

Hafner, H. (2000). Heron nest site conservation. In: Heron conservation (eds. A. J. Kushlan and H. Hafner). Academic Press, U.S.A. pp. 201-215.

Hafner, H. and Britton, R. (1983). Changes of foraging sites by nesting Little Egrets *Egretta garzetta* L.in relation to food supply. Waterbirds 6: 24-30.

Hafner, H. and Fasola, M. (1992). The relationship between feeding habitat and colonial nesting Ardeidae. In: Managing Mediterranean Wetlands and their Birds (eds. H. Hafner, M. Fasola) IWRB Press, London. Pp. 194-201.

Haines, H. H. (1924). *The Botany of Bihar and Orisa-VI.* Allahabad: Central Book Depot.

Hooker, J. D. (1872-1897). *Flora of British India.* 1-7. London: L. Reeve.

Hora, S. L. (1921). On some new record and rare species of fish from Eastern Himalayas. *Records of Indian Museum,* 22(5): 731-744.

Hora, S. L. (1937). Notes on fishes in the Indian Museum. XXXV. A further note on Hamilton's *Cyprinus (Garra) lamta. Records of Indian Museum,* 29: 344- 348.

Hoyer, M. V. and Canfield Jr., D. E. (1990). Limnological factors influencing bird abundance and species richness on Florida lakes. *Lake Reserv. Mgmt.,* 6: 132-141.

Jayaram, K. C. (1999). *The Freshwater Fishes of the Indian Region.* Delhi. Narendra Publishing House.

Kazantzidis, S.; Gunter, V.; Pyrovetsi, M. and Sinis, A. (1997). Comparative nest site selection and breeding success in 2 sympatric Ardieds, Black-crowned Night Heron *Nycticorax*

nycticorax and Little Egret *Egretta garzetta* in Axios Delta., Mecidonia, Greece. Colonial Waterbirds. 20: 505-517.

Khedkar, K. K. and Dixit, A. J. (2003). Physico-chemical analysis of domestic wastewater of Amravati (Maharashtra). *J. Aquat. Biol.*, 18(1): 69-72.

Laaksonen, T., Ahola, M., Eeva, T., Vaisanen, R. A. and Lehikoinen, E. (2006). Climate change, migratory connectivity and changes in laying date and clutck size of the pied flycatcher. Oikos, 114: 277-290.

Lack, D. (1968). Ecological adaptations for breeding in birds. Chapman and Hall, London.

Ludwig, E.; Vanicsek, L.; Torok, J. and Csorgo, T. (1994). The effect of nest height on the seasonal pattern of breeding siccess in blackbirds *Tardus merula*. Ardea. 83: 411-418.

Madsen, J. and Fox, A. D. (1995). Impacts of hunting on waterbirds: a review. Wildlife Biology, 1, 193–207.

Manna, C. K., Pramanik, A.K., Santra, K. B. (2019). Breeding behaviour and parental care of Open-billed Strok (1st Ed.). Akinik Publications, New Delhi, India.

Menon, A. G. K. (1999). Check list-fresh water fishes of India. Records of Zoological survey of India. *Miscellaneous Publication, Occasional Paper*, 175: 1-366.

Mertens, R. (1974). Die amphibian and reptilian West Pakistan. *Senckenb Boil*, 55(1-3): 35-38.

Mohsin, A. B. M., Haque, E. (2009). Diversity of Fishes of Mahananda River at Chapai Nawabgonj District. *Research Journal of Biological Sciences*, 4(7): 828- 831.

Munawar, M. (1970). Limnological studies on freshwater ponds of Hyderabad, India. II. The biocenose distribution of unicellular and colonial phytoplankton in polluted and unpolluted environments. *Hydrobiologia*, 36(1): 105 - 128.

Murray, J. A. (1884). *The Vertebrate Zoology of Sind*. London and Bombay.

Naher, H.; Sarkar, N. J.; Rahman, M. K.; Imam, S. and Zia, S. (2009). Breeding biology of the Little Cormorant *Phalacrocorax niger* (Pelecaniformes: Hlacrocoracidae) in Bangladesh. Journal of Threatened Taxa. 1(4): 221-225.

Naugle, D. E.; Johnson, R. R.; Meeks, W. A. and Higgins, K. F. (1996). Colonization and growth of a mixed-species heroney in South Dakota. Colonial Waterbirds. 19: 199-206

North, M. R. and Ryan, M. R. (1989). Characteristics of lakes and nest sites used by yellow-billed loons in Arctic Alaska. J. Field Ornithol. 60: 296 - 304.

Patankar, P., Desai, I., Shinde, K. and Suresh, B. (2007). Ecology and breeding biology of the Cattle Egret *Bubulcus ibis* in an industrial area at Vadodara , Gujrat. Zoos' Print Journal 22(11): 2885 – 2888.

Patra, A. (2005). Studies on some aspects of ecology with special emphasis on avifaunal diversity of Santragachi and Joypur jheel,Dist.Howrah,W.B., India. University of Kalyani, Nadia, West Bengal, India. (Ph.D.thesis).

Pramanik, A. K.; Santra, K. B. and Manna, C. K. (2009). Nest-building Behaviour of the Asian Open Billed Stork *Anastomus oscitans*, in the Kulik Bird Sanctuary, Raiganj, India. Our Nature. 7: 39-47

Pramanik, A. K.; Santra, K. B. and Manna, C. K. (2010). Abundance and diversity of plants and animals in the Kulik Bird Sanctuary, Raiganj, West Bengal, India. Journal of Biodiversity. 1 (1): 13-17.

Prasad, B. (Ed.) (1962). *The Wealth of India*. Vol. IV. Fish and Fisheries. New Delhi: CSIR.

Saijo, Y., Mitamura, O., Hino, K., Ikusima, I., Matsumura, T. and Silva, V. P. (1997). Physico-chemical features of rivers and lakes in Pantanal wetland. *Jap. J. Limnol.*, 58(1): 69 - 82.

Sammaiah, C. and Singh, J. L. (2004). Species diversity of water birds as a pollution indicator. Proc. Int. Con. on Bird and Environment, Haridwar, India. p. 59.

Sanjit, L. and Bhatt, D. (2004). Status and current threats to avian biodiversity of the Lokta Lake in Manipur. Proc. Int. Con. on Bird and Environment, Haridwar, India. p. 129.

Sen, N. (2000). Occurrence, Distribution and Status of Diversified Fish Fauna of North East India. In: AG Ponniah, UK Sarkar (Eds.): *Fish Biodiversity of North East India*. Lucknow: NBFGR, NATP Publication, pp. 31-48.

Sen, T. K. 1985. The Fish Fauna of Assam and Neighbouring Northeastern States of India. Records of Zoological survey of India. *Occasional Paper*, 64: 1-216.

Sharma, A. (2007). The Asian open billed strok Thrives in the Kulik Bird. The Statesman, Daily, India, August 07, 2007.

Smith, M. A. (1935). *The Fauna of British India, Including Ceylon and Burma*. Reptilia and Amphibia. Volume II: Sauria. London: Taylor and Francis Ltd.

Smith, M. A. (1943). *The Fauna of British India, Including Ceylon and Burma*. Reptilia and Amphibia. Volume III: Serpentes. London: Taylor and Francis Ltd.

Smith, V. R. (1978). Animal-plant- soil nutrient relationships on Marion Island (Subantarctic.), Oecologia, 32: 239-253.

Soler, J. J.; Moller, A. P. and Soler, M. (1998). Nest building, sexual selection and parental investment. Evolutionary Ecology. 12: 427-441.

Stevens, M. A. and Boness, D. J. (2003). Influences of habitat features and human disturbance on use of breeding sites by a declining population of Southern Fur Seals (*Arctocephalus australis*). Journal of Zoology 260: 145-152.

Subba Rao, N. V. (1989). *Handbook of Freshwater Molluscs of India*. Calcutta: Zoological Survey of India.

Subramanya, S. (1996). Distribution, status and conservation of Indian herons. Journal of Bombay Natural Society. 93: 459-486.

Surana, R., Subba, B. R., Limbu, K. P. (2007). Avian diversity during rehabilitation stage of Chimdi Lake, Sunsari, Nepal. *Our Nature*, 5: 75-80.

Talwar, P. K., Jhingram, A. G. (1991). *Inland fishes of the India and adjacent countries*. New Delhi. Oxford and IBH Publishing Co.

Thomson, A. L. (1950). Factors determining the breeding season of birds: an introductory review. Ibis 92: 173-184 (cited in Farner and King, 1971).

Tuite, C. H., Hanson, P. R. and Owen, M. (1984). Some ecological factors affecting winter waterfowl distribution on inland waters in England and Wales, and the influence of water based recreation. J. Appl. Ecol. 21: 41-62.

Welch, P. S. (1952). *Limnology* (2nd Edition). Mc.Graw–Hill Book Co., New York.

♣

Other Related Books

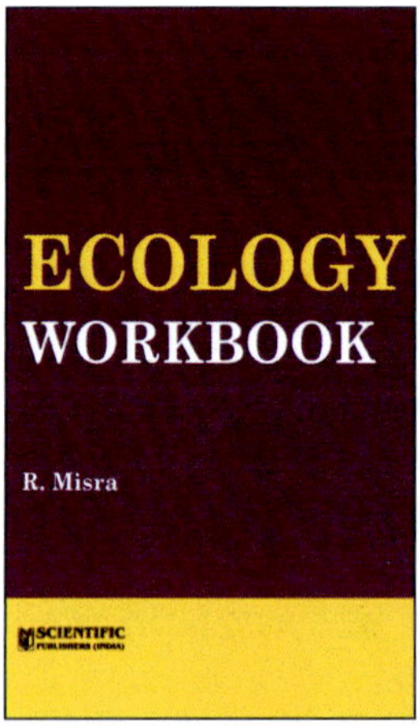

ISBN: 978-81-7233-482-6 ISBN: 978-81-7233-805-3 ISBN: 978-93-83692-27-9

ISBN: 978-81-7233-974-6 ISBN: 978-93-89412-39-0 ISBN: 978-93-89832-06-8

www.scientificpubonline.com